D0906506

ASPCA® kids

bringing friends together, paw in hand

Amazing Pet Tricks

Kate Eldredge
Jacque Lynn Schultz, CPDT

**With the ASPCA's Carmen Buitrago, CPDT; Kristen Collins, CPDT;
Dr. Emily Weiss, CAAB; and Victoria Wells, CPDT;
each of whom contributed one or more tricks to this book**

WILEY

Wiley Publishing, Inc.

Howell Book House
Published by Wiley Publishing, Inc., Hoboken, New Jersey

Photo credits: All photos ©jeanmfogle.com except on the following pages: 17: ©iStockphoto.com/Michael Westhoff. 27: ©iStockphoto.com/Jean Frooms. 47: ©iStockphoto.com/Serega. 62: ©iStockphoto.com/Tatiana Boyle. 63: ©iStockphoto.com/James Cote. 66: ©iStockphoto.com/Matthew Gough. 71: ©iStockphoto.com/Mark Hayes. 75: ©iStockphoto.com/Vasiliki Varvaki. 77: ©iStockphoto.com/macatack. 80, 87: ©iStockphoto.com/Jill Lang. 83: ©iStockphoto.com/Greg Henry. 84: ©iStockphoto.com/Serdar Yagci. 86: ©iStockphoto.com/smiley joanne. 88: Annette Wiechmann.

The publisher and the author make no representations or warranties with respect to the accuracy or completeness of the contents of this work and specifically disclaim all warranties, including without limitation warranties of fitness for a particular purpose. No warranty may be created or extended by sales or promotional materials. The advice and strategies contained herein may not be suitable for every situation. This work is sold with the understanding that the publisher is not engaged in rendering legal, accounting, or other professional services. If professional assistance is required, the services of a competent professional person should be sought. Neither the publisher nor the author shall be liable for damages arising here from. The fact that an organization or Website is referred to in this work as a citation and/or a potential source of further information does not mean that the author or the publisher endorses the information the organization or Website may provide or recommendations it may make. Further, readers should be aware that Internet Websites listed in this work may have changed or disappeared between when this work was written and when it is read.

For general information on our other products and services or to obtain technical support please contact our Customer Care Department within the U.S. at (800) 762-2974, outside the U.S. at (317) 572-3993 or fax (317) 572-4002.

Wiley also publishes its books in a variety of electronic formats. Some content that appears in print may not be available in electronic books. For more information about Wiley products, please visit our web site at www.wiley.com.

Library of Congress Cataloging-in-Publication Data:
Eldredge, Kate.
   Amazing pet tricks / Kate Eldredge, Jacque Lynn Schultz.
      p. cm. — (ASPCA kids)
   Includes index.
   ISBN 978-0-470-41083-7
   1. Pets—Training—Juvenile literature. I. Schultz, Jacque Lynn. II. Title.

SF412.7E43 2009
636.08'35—dc22

2008046485

Printed in China

10 9 8 7 6 5 4 3 2 1

Book design by Erin Zeltner
Book production by Wiley Publishing, Inc. Composition Services

*This book is dedicated to Flash, my first
dog and most dedicated trickster, and to Zoom,
the goat who loves to bow.*

Acknowledgments

I would like to thank my mom, Deb Eldredge, for encouraging me to write and pursue my love of animals, as well as letting me steal the computer to write late at night. I would also like to thank my father, Chuck Eldredge, who fixed the aforementioned computer whenever it decided it didn't want to play the game. Thanks to my brother, Tom, for making the occasional batch of popcorn, though you did grumble about it.

Then a big thanks to the hodgepodge of critters who tried out and learned various tricks: the dogs, Flash, Tia, and Queezle; Frodo the mini-horse; Sugar the donkey; and Zoom the goat; plus many others, including Baloo who decided that since B stands for *Baloo*, no other letter matters. Special thanks, also, to my mom's dog, Hokey, who never let me forget that dinner was at five and bedtime chews at nine—and make it sharp!

A big thank you to the ASPCA and Marion Lane for requesting me to write this book. And thanks, of course, to my editor, Pam Mourouzis; my agent, Jessica Faust; and everyone else at Wiley for helping to make this work.

Contents

To the degree that we come to understand other organisms we will place greater value on them and on ourselves.

—E.O. Wilson

We cannot do great things on this earth. We can only do small things with great love.

—Mother Teresa

My enjoyment of animals began with Happy, my first dog. She was energetic, loving, fun, and always available for anything—from racing me around the yard to lying quietly by my side while I finished my homework.

But it was the bird feeder at my window that opened up the mystery of it all. From just a few feet away I could watch these alert, colorful creatures nourish themselves while keeping a wary eye on everything around them. I had so many questions about the birds: What did they do all day? Where did they go at night? What happened to them during storms? I thought there ought to be a daily newspaper to report on what had happened the night before in my yard. Who had survived? Who hadn't?

These questions inspired me to observe more and read more about birds. Then I began to learn about other animals—chimpanzees, snakes, and whales—by reading books written by famous authors who spent their lives studying them. Jane Goodall's writing on chimpanzees was the most fascinating to me. But everything I read led to more questions!

My early interest in animals continued to grow. After college, I became involved in creating programs that teach people to care about animals, and then became director of an animal shelter in New Jersey. Other positions followed, and eventually I became president of the ASPCA—which stands for the American Society for the Prevention of Cruelty to Animals—the first humane organization in the United States. My childhood curiosity led to a career helping animals, and that enriches my life beyond description.

This past year, the 400-member ASPCA team not only saved the lives of thousands of animals in this country, but also enforced the laws designed by our society to protect animals. The mission of the ASPCA from 1866 until the present day is "to provide effective means for the prevention of cruelty to animals." One way the ASPCA does this is through education—through written materials for young people. The books that I read when I was young fed my growing interest in animals. I hope this book will do the same for you.

Edwin J. Sayres
ASPCA President & CEO

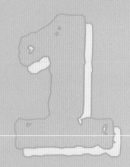

Getting Started

So you want to teach your animal friend to do tricks. Congratulations! As you have probably already guessed, tricks aren't just for dogs. With a little bit of work and patience, cats, horses, birds, and even turtles can all learn to do fun and amazing tricks. Different animals may require slightly different training techniques, but all of them can join in the fun. Before you get started, though, there are a few things you should know.

Treats

Most animals are pretty much food motivated, meaning they really like food and will do many different things to earn a treat. Dogs are the most obvious about it, as they sit beside the dinner table and beg, but other animals like food a lot, too. This love of food can be very useful for trick training.

The ideal training treat is small and soft so that your pet won't have to do a lot of chewing. Also make sure that it is something safe for your pet to eat and something he really likes. This will make him even more eager to work for the treat. Less exciting treats, such as dry kibble, often only work well if your pet hasn't eaten in awhile.

For dogs, you can use anything from bits of kibble to small chunks of cheese or deli meat. Look around in the dog food aisle of your local pet supply store or grocery store and you'll find all sorts of different dog treats. Some dogs prefer tastier treats such as cold cuts, while others are less picky and will settle for biscuits. Find out what your dog really likes to eat before starting, because you won't get very far if Prince Charming feels he deserves something better than what you're offering.

If your dog has food allergies, make sure to offer treats that don't have the ingredient he is allergic to. The wide variety of dog treats makes it very likely that there will be something he can have.

If you are training a cat, try out some cat treats. Cats are often pickier than dogs, so don't buy a whole bunch of one type of treat until you're sure your cat will like it. Again, you can also use small bits of deli meat or some of your cat's regular dry cat food.

Farm animals such as horses and goats are really easy to find treats for, and most likely you have some in your house. Sure, you can get special horse treats, but most horses are just as happy to get an apple or a carrot. My goat likes Cheerios. Other types of cereal could also work for sheep and goats. Just avoid brands with a ton of sugar. If you're using apples or carrots, be sure to chop them up into bite-size pieces so that you won't need a whole bushel of apples for one training session.

You should be able to find treats for almost any type of animal in your local pet supply store. Otherwise, just try out different foods. Be sure that they won't harm your pet before you let him eat anything new. Small nuts work well for birds, and turtles enjoy fruits and veggies. Check out the ASPCA's Animal Poison Control Center (www.aspca.org/apcc) for lists of foods that are poisonous to your pet.

If your pet is overweight, you can still use treats for training. However, you must make sure that he gets less supper on days that he gets a bunch of treats. You can even make him work for his supper by having him do tricks to earn each bite. This technique can work really well for animals who aren't quite as excited about learning new things but love dinnertime. They will quickly catch on that goodies come only after they have practiced some tricks. This also makes feeding time a little more exciting for everyone involved. One of my dogs will start doing tricks on her own if she's really hungry and wants her supper!

Also, just because your pet is at an ideal weight right now doesn't mean that he should get a ton of treats in addition to his normal meals. Reduce the amount of food he gets at each meal when you are doing a lot of training so that he eats the same amount every day. This way, he will remain at a healthy weight.

If your pet doesn't like food that much, don't be discouraged. You can also use toys, play, and praise to encourage him to work and let him know when he has done something right. Rewards must reinforce the animal so he will repeat the behavior, so use whatever your pet likes best.

Basic Skills

Many tricks for all types of animals are based on basic obedience skills for dogs. Depending on the type of animal you are working with, these skills may or may not be practical. However, dogs in particular should know these four things, and cats should, too. (Despite what you may have heard, you can teach a cat just about any trick that you can teach a dog.)

Sit

Sit is the champion of all skills. A lot of tricks, such as shake, wave, and high-five, require your pet to know how to sit on cue.

Start off by getting some treats ready. Make sure they are cut into small pieces, and have them in a convenient location. Some good places are in your pocket, a fanny pack, or a special dog treat bag called a *bait bag* that you can get at a pet supply store. Or you can place treats on a nearby table or other raised surface so that your pet can't steal them. Hold one treat in your hand, and stand with your pet toe-to-toe in front of you or on your left side (which is the traditional heel position for dogs).

Start by holding the treat right in front of your pet's nose. Then slowly bring it up and back toward his tail. As he tips his head back to reach the food, he will most likely sit. When he does sit, say "yes" or "good" and give him the treat. This technique is called *luring*. That means using a treat to lead your pet into the position you want. Kind of like fishing, the treat is your lure. Only instead of catching a fish, you use it to lead your animal friend into the correct position.

After several repetitions, your pet should begin to expect what you're going to ask him to do and sit without needing to go through the entire exercise. At this point, you can start saying "sit" before moving your hand because your pet has shown that he understands the behavior you want. Repeat several times, and begin to reward him only when he sits after you ask him to.

Dogs and cats are pretty smart. Before long your pet will put the pieces together and sit without help when he hears the word "sit" so he can get his treat faster.

If your pet backs up instead of sitting when you use the treat, try either starting with his rear end against a wall or just blocking his rear with your hand so that he can't back up. As he gets the hang of it, he will no longer need the hand or wall. If he jumps up to get the treat, try holding your hand lower down.

Down

There are two ways to teach your pet to lie down. Either way, you will need to have your treats ready just like you did when you taught him to sit.

One way is to start with your pet sitting. Take a treat in your hand and hold it in front of his nose. Slowly bring the treat down to the floor and then either forward or back—whichever way causes your pet to drop his elbows and lie down. When he does, say "good!" and give him the treat.

Once your pet is beginning to anticipate your hand motion and doesn't need as much help going into the down position, you can add your cue word. Say "down," then move your hand. At this point you can begin to fade out (gradually stop using) your hand motion. Eventually, your animal companion will lie down when he hears your cue. Only reward him for a down that *you* asked for to reinforce the idea that he is responding to your cue.

The other method is to start with your pet standing. You may hear this referred to as a *fold-back down* because your pet will fold back into the down position like a folding chair. This is the best choice if you hope to go on to compete in formal obedience competitions with your dog. Also, some tricks may require your pet to lie down without sitting first.

Begin by holding a treat in your hand just in front of the pet's nose. Then move your hand down and back between his front legs. This should cause him to fold his legs and drop down to get the treat. If need be, apply *gentle* pressure between the shoulder blades to encourage him to lie down. When he lies down, be sure to say "good!" and give him his treat.

Come

Start out in a small enclosed area. Be sure to have your treats with you! When your pet isn't paying attention to you, call his name

and say "come!" while running backward away from him. Your pet, seeing motion and wondering what on earth you are doing, will come running toward you. When he does, say "good!" and give him a treat while touching his collar. Then release him and repeat the process. After a while, he'll come without you needing to run because he'll understand that he gets a treat when you call him.

As your pet comes to you more reliably, you will need to practice in areas with more distractions. With a dog, it could be a neighbor's yard (get permission first!) or a local shop that allows dogs to come inside. It can be very hard for dogs to leave something that smells good, so be patient when you practice in a new area. For a cat, try working around distractions like an aquarium or unfamiliar objects that he will be interested in. The idea is to get your pet used to focusing and paying attention to you even if there are other things around that he would like to do or check out. Be patient! Always go back to step one when you practice in new locations.

Although come isn't very hard to teach, it can be very hard to get your pet to do it all the time. There are a few tricks to this. First, don't call your pet only when you want him to stop doing something fun or get in the car to go to the vet. He will quickly figure out that when you call, something bad happens or the fun ends.

To prevent this, call your pet often when he is playing and having a good time. When he comes, praise him and give him a treat. Be sure to grab hold of his collar and hold him for a few seconds. Then release him and allow him to go back and play some more. This way, he never knows if your call is going to end playtime or not.

The second key is to call your pet only when you are able to go get him if he doesn't respond. Otherwise, he will learn that he doesn't have to do what you say. Someday your pet might be in danger, and you will need him to come the first time you call. If you get into the habit of calling him four or five times before he comes, you could end up with a problem.

If your pet is investigating an interesting smell and doesn't come when you call him, go get him and lead him quietly inside or to another room, ending the fun. This correction doesn't require any yelling—just go get him and give him a little quiet time. Then take him out and try again in a less interesting place so he isn't so distracted. Make a big deal of it when he comes.

For a pet who has a hard time with the come cue, always have treats with you so that you can reward him when he comes the first time you call. He will figure out that coming when called means he gets a treat and gets to continue playing, and that not coming on the first call means the fun ends.

Stay

Teaching a pet to stay is not always easy, but it is very useful. Your pet will need to know how to sit or lie down first. Begin by putting him in a sit or a down. Once he is in position, say "stay" and put your open palm in front of his face. This will be your stay signal. Wait just a few seconds, and if he hasn't moved (you shouldn't give him enough time to), say "good!" and give him a treat while he is still in position.

You can't ask your pet to stay forever. He needs to know that *stay* means "stay there until I tell you that you don't have to stay anymore." You'll use a word such as "okay!" to let him know that he can get up and move around. This is called a *release word*.

To practice the stay, ask your pet to sit or lie down, give the "stay" cue, wait a few seconds, and then give him a treat and say "okay!" Gradually increase the time he has to stay in position to get the treat.

When he gets good at this, don't always release him right after he gets the treat. Have him stay for a few seconds, give him a treat, and then have him stay a little longer before he gets another treat and is released. If he has trouble, go back to a shorter length of time. Remember that this will take awhile for him to learn, and be patient.

Once your pet is beginning to get the idea, you can start adding distance. Start by going only a step or two away for a short amount of time. As he grows more confident, you can increase the distance you step away from him. Eventually you will even be able to go around corners and out of sight without him getting up—although that takes a lot of time and practice.

If your pet gets up before you release him, quietly lead him back to where he is supposed to be and put him back in position. Then repeat the cue "stay" and try again for a shorter time. If your pet is staying but looks like he might get up, try reminding him by repeating "stay" in a quiet voice. Hearing the cue again may remind him of what he's supposed to be doing.

To get a reliable stay, you will need to practice in many different places. For dogs and cats, this means practicing both outdoors and indoors—especially with dogs. When you're working outdoors, unless your pet is 100 percent reliable on the come cue, keep him on leash or in a fenced enclosure.

Be patient when you and your pet are working near new distractions. Most pets have trouble realizing that an old cue in a new situation is really just the same old thing. Your pet may need help remembering what you want him to do. Go back to short times and distances, and work your way up in each new place. You will be able to increase your time and distance more quickly because your pet already knows what he should do. He just needs to understand how to apply it to a different setting.

Some Things to Remember About Basic Training

You can use any word you want as a cue for different behaviors. Some people train their pets using a foreign language, or you can use something silly like "tulip" for down. Just be sure *you* know what word you want to use and stick with it. Make sure everyone in the family commits to using the same cues. Consistency—doing things the same way every time—is very important. You may find that the traditional words are easiest because you think of them first.

Also be careful about mixing cues. For example, don't tell your pet "sit down." Most likely he will look confused, wondering what you want! It's better to use just plain "sit" or just "down," depending on which position you are actually looking for.

When giving cues, you should remember a few basic things.

- Don't give your pet a cue if he isn't paying attention—it won't get you anywhere. If your pet isn't looking at you, say his name first to get his attention.

- Use a clear, upbeat voice, and make sure that your pet can hear you. But you don't need to yell. Shouting will just make him feel like he has done something wrong, and he won't want to work with you.

- Try to stick to one- or two-syllable cue words because they are easier to say and remember. You can dress up some tricks by sticking your cue word into a sentence or question, but your pet needs to know only the one word that tells him what to do.

- Introduce your cue word only after your pet knows the behavior you are asking for.

- Don't ask your pet to do something in a situation where you can't help him out if he doesn't do it. For example, don't call your pet to come if you can't go and get him if he doesn't come. This just shows him that he doesn't have to do what you ask and will cause problems later on.

- Reward your pet only if he does what you ask on the first cue! You should not have to repeat your cue several times to get him to do what you want.

This sounds like a lot of "don'ts," but they will help you and your pet to be successful tricksters.

One last thing to remember: Learning new skills takes time, so you will need to be very patient. Remember when you learned how to read? First you had to learn the alphabet, and then you moved on to small words and gradually to full sentences. All in all it took a lot of time and practice. Training your pet is just like learning to read. You will need to practice, practice, practice, and it will take a bit of time to sink in. Just be patient and keep at it, and eventually your hard work will pay off.

2 The Trick Is in the Click

J ust about all the tricks in this book use a system called clicker training. A clicker is a little colored box with a button that makes a "click" sound when pressed. (Many pet supply stores carry clickers. So do stores that sell party favors.) It's a useful tool for teaching tricks and can be a lot of fun to use. Clicker training was originally used to teach dolphins, but now it's used to teach all kinds of animals many different skills.

The way clicker training works is pretty simple. First you teach your animal companion that whenever she hears the click, she has done something good and a treat is on the way. Once she understands that idea, you can use the clicker to encourage her to try new things and figure out new behaviors by herself. When she does what you want her to do, you can click and then give her a treat. Pretty soon, she'll be working hard to figure out what you want so she can earn that click and treat.

Using a clicker is kind of like saying "good!" when you teach the basic skills in chapter 1. It's a way to communicate to your pet, "Yes, that's just what I want you to do!" So why not just use your voice? Because, being human, our timing is often a little off. Using a clicker forces us to think about the exact moment our animal companion is doing something we like, and the sound of the clicker marks the moment of the behavior. The click is also a noise that your animal companion won't hear all the time in everyday life, and she will soon associate it with learning new things. However, if you have trouble juggling treats, clicker, a prop, and maybe even a leash, feel free to use your voice instead of the clicker. You will still get good results.

Introducing the Clicker

Before you can do any serious training, you need to get your pet used to the sound of the clicker and teach her what it means. Have some treats handy. When you are ready, click the clicker and give her a treat. Right now, she doesn't have to do anything for the treat. Just click and treat, click and treat. Repeat this several times in a row.

Now try waiting until your pet has lost interest in you and is sniffing or looking at something else. Then click. She should turn and run to you to get the treat. If she doesn't, go to her and give her a treat and repeat the first step a few more times. Make sure she gets a treat every time you click—even if you have to go over and give it to her. The goal is for her to get really excited when she hears the click.

Once your pet understands how the clicker works and looks to you whenever she hears the sound, you can try it out on some behaviors she already knows. This will help her understand that the click means the same thing as "good girl!" So, for example, if you've already taught your pet to sit, tell her "sit." When she does, click and give her a treat. Try it a few times with each of the skills she already knows. This teaches her that when she does what you want her to do, she can "make you" click and give her a treat.

Teaching Something New

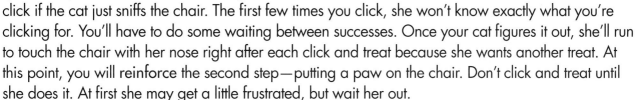

The goal is for your pet to figure out for herself what you want her to do. In many ways this is like the game Hot or Cold. She'll try different behaviors because she wants you to click and give her a treat. When she does what you want, you click. Otherwise, you do nothing. Using a clicker puts the responsibility on your pet and reinforces her for using her mind and trying to solve problems.

The key is to remember that your pet won't be able to do the whole behavior right away, so you need to reinforce all her baby steps on the way to your final goal. For example, let's say you're teaching your cat to sit in a chair. With treats and clicker ready, set up the chair in your training space. Be patient and wait for her to go to the chair on her own. At first, click if the cat just sniffs the chair. The first few times you click, she won't know exactly what you're clicking for. You'll have to do some waiting between successes. Once your cat figures it out, she'll run to touch the chair with her nose right after each click and treat because she wants another treat. At this point, you will reinforce the second step—putting a paw on the chair. Don't click and treat until she does it. At first she may get a little frustrated, but wait her out.

When she's putting one paw on the chair, wait until she has two paws on the chair before you click and treat. Eventually, she will jump all the way into the chair—without you having to do anything but click and dish out treats!

This process of reinforcing each little step of a new skill is called *shaping*. With an end behavior in mind, you can reinforce behaviors that are related to but not quite what you want, requiring your pet to get closer and closer to the finished behavior to earn a treat. When your pet really knows a step, you can ask for something a little closer to what you actually want.

Introducing a Cue

When you're teaching your pet new tricks and skills, you will eventually need to use some sort of cue, or signal, so that she can tell which trick you are asking for. Work on only one trick per training session. Each session should be just three to ten minutes long.

First, teach your pet the behavior you want her to know. For example, let's say you're teaching her to sit. Once she knows how to sit on cue and is doing it consistently to get a treat, you can introduce your cue word. After she has performed a bunch of good sits to get you to click, wait a moment and say "sit." When she does it, give her a treat. At first the word won't mean anything to her. Keep repeating it when you want her to sit or when you see her start to sit, and she will begin to associate the word with her action. At this point, reward her only if she sits when you tell her to. With some practice, she should be sitting on cue.

Let Your Pet Be the Driver

Try carrying a clicker and treats around in your pocket during the day. Many dogs do cute or funny little things on their own, like leaping in the air and spinning when they are out for a run. You can click when they do this, and then turn the natural behavior into a trick that you can request on cue. For example, when your dog leaps up and spins, you can say "spin" and click the clicker, then give her a treat for doing something you liked. This type of training is called *capturing a behavior.*

Having the clicker and treats in your pocket all the time is also a good way to encourage good behavior. For example, when you and your dog are out on a walk, you can click and treat when she isn't pulling on the leash. Obviously, don't click if she is pulling. This makes walking nicely by your side more rewarding than pulling, so she will pull less. This can make a walk around town much more enjoyable.

Alternate Routes

Some animals are scared of the sharp click sound or just don't like it. If this is the case, you can use a word, such as "yes" or "good," instead of the click to mark the correct response. There is no difference between using a click or a word, except that you use a clicker only when you are training. So your pet knows the click sound is always about training, while your voice is sometimes not directed at her at all. Still, the clicker principles work even if you are using a verbal marker—a word such as "good!" Another option is to use something that makes a softer or quieter sound, like a clicky pen.

If you are the type of person who has trouble doing two things at once, handling a clicker, treats, and possibly a leash may seem daunting. In that case, just use a

verbal marker to make life easier for you. It is more important for your timing to be good with a verbal marker than with the clicker. You want your animal companion to know the exact moment when she has done the right thing.

Another situation when you might want to use a verbal marker instead of a click is if your pet gets overexcited when working with the clicker. This enthusiasm is good, but for some tricks it can work against you. In those cases, put the clicker away and use a verbal marker until you are doing something where you can put your pet's energy to good use.

Clicker Games

One fun way to try out clicker training and play around with how it works is to train a person. Find a willing participant, maybe a sibling or a parent or friend, and explain to them that you are going to try to teach them to do something. They have to guess what it is, and whenever you click, it means they're on the right track.

Next, decide what you are going to have them do. Pick something fairly simple, like sitting in a nearby chair. Let them know when you have decided, but don't say what it is you want them to do.

At first you will need to click if they take a step toward the chair. If they go in the wrong direction, don't click. Eventually they will get to the chair, guided by your clicks when they took a right step. Next, you will need to click if they touch the chair, and so on until they pull out the chair and sit in it. Once they have completed the action, let them know that they got it!

Now switch roles—your friend can be the trainer, and you can act as the subject. Experiment. Try moving in different directions and touching different objects. Whenever you hear a click, you are headed in the right direction.

The Magic Touch

A touch stick and a target are two very useful tools for teaching tricks to a wide variety of animal companions. Both are sort of a mini-trick in themselves. They'll also help you guide your animal friend in learning new tricks. You will be able to use them as tools to teach many new tricks and behaviors in the future, and they can help your pet understand new things.

The Target

Find a little plastic lid, like the top of a margarine tub or round cardboard oatmeal or breadcrumbs container. Almost anything will work as long as it is flat and fairly small, but big enough for your pet to be able to clearly see. This will be your target. Next, decide on the cue word you would like to teach your pet to associate with the target. "Target" and "touch" are two popular cue words. Once you choose a word, be sure to use the same word every time. The target is used in a lot of dog tricks, but it works with other animals, too, including cats.

Now, gently hold your pet near the target, but make sure he can't reach it just yet. Place a treat on the target and release him. He will probably go right to the target to get the treat. This is good, because you want him to like going to the target. Repeat several times. When your pet is consistently going to the target to get the treat, start using your cue word before you release him.

The next step is to begin increasing your distance from the target. Ultimately, you want your pet to be able to go to the target even when it is far away from you. This will be useful when you're working on distance for some tricks later on.

At first, move the target a few feet from you. After a few repetitions, move it a little farther away. As the target moves farther away, you will probably need someone to help put the treat on it while you hold your pet. Get a friend or family member to put a treat on the target each time you send your pet.

You can also use this as a chance to practice your pet's stay skills. Tell him to stay, and then place the treat on the target. Be ready to stop him if he moves, and don't let him get the treat early! If he gets up, catch him and take him back to where you told him to stay. This can be very hard, as your pet has to resist the desire to eat. Be patient! Also, be sure not to release him until you are back next to him so he won't start anticipating and going for the treat early.

For pets who really can't wait, there are two solutions. One is to have a helper at the target to cover the treat if your pet takes off before he is told. This way, he can't get the treat and be rewarded for doing the wrong thing. The other is not to put any food on the target. Just send your pet, and then follow him and hand-deliver the treat at the target.

Using the Target

Mostly, you'll use the target to encourage your pet to move away from you when you're teaching him to do things at a distance. For example, if he has a really solid sit, you could put a target out, send him to it, and tell him to sit when he gets there. He should sit right there, even if the target is across the room from you.

You can also use the target to get your pet to touch or go to other objects, such as doors or fences. You can tape or tie it to fences or doors (just punch a hole in the target) to get your pet to touch them. Use this skill to teach your pet to close the door or turn off the light.

Another nifty way to put your pet's target knowledge to use is to teach him cues for specific locations, such as his bed or food bowl. By repeatedly placing the target in that location, you can then switch in the name of that place, such as "go to bed," or "go to your bowl," for your target cue. Once your pet is consistently responding to the new cue, you can remove the target and he will go to that location on your cue.

The Touch Stick

The touch stick is a useful tool for all pets, from dogs to horses to fish. All you need is a light wooden stick. At the hardware store they call it a *dowel*. There are official touch sticks, but anything you can use like a pointer is fine. It's best if it has a tip that is a different color from the main part of the stick. This makes it more visible to your pet. You can use a nontoxic marker to color the end. To increase visibility, you can put a small ball on the end of the stick. For cats, feathers work great.

Get your clicker and some treats ready. Present the end of the touch stick to your pet. When he sniffs it (licking or grabbing is okay at first, but try not to encourage it), click and treat. Then start giving him a cue, such as "touch." You can use the same cue for both the target and the touch stick, but it is better to have two different cues. I use "target" for the target and "touch" for the touch stick, because they are both easy for me to remember.

Practice in many different areas and settings. You can also teach your pet to touch your hand with his nose in the same way. That way, if you are teaching him something and need him to touch but don't have the stick with you, you can use your hand instead. After all, the stick is just a tool you use to get your pet to perform a specific behavior. The behavior can occur even without the stick.

Using the Touch Stick

You can use the touch stick as a lure to guide your pet into position. For example, by holding the end of the stick so that he has to turn to his left, you can teach him to turn left. You can also use it to get him to touch things he normally wouldn't be interested in.

If you get involved with musical freestyle, a dog sport in which people and their dogs perform routines choreographed to music, you can use the stick as a prop for your performance. You can decorate your stick with paint and even glitter to make it fit your theme for skits and dances—or just because you like it that way.

Classic Dog Tricks

Some tricks have been passed down and around for what seems like forever. Shake hands, take a bow, play dead—we just can't get enough of them! They are fun and cute, and every true trick dog has them all down pat.

There are several acceptable ways to train an animal companion. Animals have very different personalities: some are shy, some are bold, some are meek, and others are stubborn. Some like to try new things (these are great pets for clicker training), while others are afraid to and need to be gently coaxed. You can successfully train most dogs using one of these three methods:

1. Lure and reward training, using food, a toy, play, praise, or a combination of these to lure the dog and then reward her
2. Shaping a behavior using clicker training
3. Physical prompting, where the handler (that's you) pushes, pulls, or otherwise puts gentle physical pressure on the dog's hips or shoulders or behind the knees

Sometimes you have to be creative, but as long as the training is humane—that is, it doesn't cause physical harm, unnecessary discomfort, or undue distress to your pet—you are probably okay.

Shake

A lot of dogs naturally offer a paw to shake, making it a very easy trick for them to learn. For other dogs, it is more difficult. Of my three dogs, two understood shake almost immediately, while the other didn't get it at first.

Start by sitting on the floor with your dog sitting and facing you. If your dog gets pushy or silly when you get down on the floor, or thinks it's time to wrestle, sit on a chair or stool. Have your clicker ready and treats nearby.

First, hold your hand out where your dog could easily put her paw in it, and wait. Always hold your hand with your palm facing up. Many dogs immediately start trying to figure out what you want, eagerly playing the game. Your dog may touch your hand with her nose, bark, or even get up and circle you. She will probably run through every behavior she already knows to see if one of them is what you want. Just ignore all this and wait her out.

Even if it takes a while, it doesn't mean your dog isn't smart. She's just thinking and working things out for herself. With any luck, she will eventually paw at your hand.

As soon as your dog's paw touches your hand, click and treat. As always, you can verbally praise her, too. (You can wait until later to add your cue word.) Then start again.

To make the process go quicker, feel free to hold a treat in your closed fist so your dog will try to get it out. Just be sure to wait until she uses her paw, and don't let her get the treat for licking.

If you have been waiting a long time and your dog is beginning to look frustrated, make it easier. If you haven't been holding food in your hand, do so now. If she is still having trouble, place your hand on the floor close to her paw, or even tap the floor like you're playing. This may encourage her to hit your hand with her paw.

As soon as there is any contact between paw and hand, click and treat. Be sure she is touching your hand consistently before you move on and make it harder. As she gets more confident, you can gradually raise your hand up off the ground.

With the treat-in-hand method, you will need to gradually wean your dog off the food in your hand. Never reward her with the treat that's in your hand—always feed her from your pocket or a treat bag. This lowers the value of the treat in your hand in your dog's eyes, and will make it easier to stop giving her the hint.

Remember, always hold your hand with your palm facing up, even when you are making a fist. Once your dog is consistently touching your hand, try holding out your hand without a treat. Your hand should smell like treats. As soon as her paw touches your hand, click and give her a treat from your stash. If she doesn't touch your palm when there is no treat in your hand, go back a step. Try holding a small treat under your thumb, with your other fingers outstretched. Be sure not to give her the treat under your thumb! After a few repetitions, you can try again without a treat.

When you click or praise your dog, you are telling her that whatever she just did is good. By repeatedly rewarding the same behavior, you reinforce the idea that the behavior is something you want your dog to do. In her mind, she is thinking, "Okay, so when I hit his hand with my paw, I got a treat. Let's try it again!" A dog is more likely to repeat a behavior that has gotten her positive results than one that has been ignored—just as you are more likely to study hard for a test when you know it will get you a good grade.

High-Five

High-five is easy to teach if your dog already knows shake. Ever the fan of serious slang when working with my dogs, I use the cue "gimme five." If you're more proper, you can use "high-five." Get your clicker and treats ready.

If your dog already knows shake, just hold your hand out palm up. This visual signal is associated with shake, and your dog will put her paw in your hand. When she does, click and treat. To get your hand from horizontal to vertical, gradually shift the angle of your hand, making sure to click and treat every time she gets it right. It shouldn't take very long. You can also try it with your hand held vertically and have your dog lean back in the sit to hit it with her paw— although this is not a good choice for dogs with long backs or stiff joints.

If your dog doesn't know shake, start by holding out your hand in front of her. Hold a treat if needed to make your hand more tempting. At first it may take her a while to figure out what you want, but eventually she will hit your hand with her paw to try and get the treat.

To start, hold your hand fairly close to the ground where it is easier for your dog to reach. As soon as her paw touches your hand, click and treat. Don't give her the treat in your hand—give her one from a separate stash. This will make it easier to wean her off the food later on.

Once she gets confident, you can begin raising your hand. Then you can also start weaning her off the food in your hand. At first, hold only a tiny treat under your thumb and then no treat at all. When she consistently hits your hand with her paw at the height you want, you can begin using your cue word.

For really exciting occasions, you can also teach your dog to hit your hand with both paws—"gimme ten!" To get your dog to use both paws at once, you will need to go back to the beginning. At first she might get frustrated because what worked before no longer works. But eventually she will get frustrated enough to pounce on your hand with both paws, at which point you click and treat. Always remember that patience is the golden rule!

Roll Over

Roll over is a trick I find easiest to teach without a clicker, simply because most dogs don't naturally roll over much and you might have to do *a lot* of waiting. Remember, with the clicker you have to wait for your animal friend to make the first move so that you can reward the behavior you are looking for. If you have a dog who likes to roll a lot, feel free to follow her around with a clicker and click and treat every time she does it.

Also, be aware that rolling over is easier for small dogs and dogs with a rounder body shape. For example, my Corgi picked it up right away and flips into a roll with no problem. My Belgian Tervuren, on the other hand, has a little more trouble and sometimes gets "stuck."

To teach your dog to roll over, start with her in the down position (lying on her belly with her elbows on the ground). Then take a treat in your hand. Dogs' bodies tend to follow their heads, and you can use this to your advantage. Starting with your hand close to your dog's nose, lure her head around and back toward her rear. At the same time, begin to move the lure over her back. This should cause her to flip over onto her side. Continue to lure her head in the direction she is rolling to get her to roll onto her back and then upright on the other side to complete the roll. At that point, praise her for doing such a good job and give her a treat.

Warning: Roll over will take a lot of repetitions and patience on your part. The more you do it, the more your dog will begin to understand the trick. As she gets it, she'll start to anticipate the roll and won't need as much help. At this point, you can introduce the cue you've chosen, such as "roll over." Before long, your dog will hear your cue and roll on her own.

If your dog already knows shake, this is an easy trick to teach. Since I usually have my dogs use their right paws for shake, I do wave with the left paw. This way, they don't get confused. Have your clicker and treats ready!

Begin by holding out your hand as if you want your dog to shake. This is a *hand signal,* a motion that your dog will associate with a particular behavior. It's much like the signals a baseball catcher uses to communicate with the pitcher. As soon as your dog raises her paw even a millimeter off the ground, click and treat. As she gets more confident, require her to raise her paw a little higher before you click. Initially, it doesn't matter which paw she raises—you can be particular about that later. Remember that the game is her figuring out what she has to do to make you click (and treat!).

Once she is consistently lifting a paw, you can add the cue "wave" along with the hand signal. You can also get pickier. If you want her to raise her left paw when you say "wave," click only when she raises her left paw. Even if she has been raising her right paw all along, if you wait her out she will eventually lift the left paw. At this point, click and treat and have a big celebration.

When you're working on wave, be sure to occasionally review shake. That's because at first your dog may get confused and have trouble understanding that these two very similar things are actually separate tricks. Practice will help her sort it all out.

Kiss

As with most tricks, some dogs are naturals. Giving a kiss on cue can really bring out your dog's personality, whether she is a shameless smoocher or a bit more shy. Some dogs look downright embarrassed when asked to kiss in public! While one of my dogs gives big, sloppy kisses to anyone who asks (or doesn't ask), another used to give only quick pokes with his nose, and half the time he didn't even want to do that.

If your dog likes to give kisses, you'll have an easy time getting her to do it on cue. Just wait until she plants one on your face, then click and treat. After being rewarded a few times, she will be more than happy to oblige.

For more reserved canines, a slightly sneaky approach is required. Have a clicker and treats ready, as well as a jar of peanut butter, some squeeze-cheese, or a flavored paste made for dogs. Start by putting a dab of the treat on your hand. You only need a little! Then encourage your dog to lick it off. When she does, click and give her a training treat as well. Repeat several times and introduce your cue. I use "gimme a kiss."

Once your dog is licking your hand regularly, switch to putting the food on your cheek. When she shows you that she understands the trick, begin to wean her off the food. After all, people don't usually walk around with peanut butter on their cheeks. Make smaller and smaller dabs, always rewarding her from your separate treat stash when she kisses you. Eventually, you won't have anything on your cheek, and she will still kiss you.

If, at any point, she starts to have trouble or act confused, just backtrack a step and use a larger dab of peanut butter, cheese, or paste to make it easier.

Bow

Always a crowd pleaser, bow is cute and classy at the same time. You can even switch it in as a greeting for more distinguished guests if you don't think they'd like to shake a canine paw.

For teaching bow, I like to use a touch stick and a clicker. That's because you can use the touch stick to guide your dog into the correct position without using your hand. This makes it easier to eventually fade away the guide and have your dog work from just a verbal cue—after she knows the trick, of course. You'll use the clicker to mark when she does the correct behavior.

Beforehand, think about what cue you want to use. Although some dogs don't have any trouble, mine seem to get a little confused because "bow" sounds a lot like "down." Another possibility is "ta-da!" like a grand finale. My Australian Shepherd takes her cue from the word "manners" when I ask her, "Where are your manners?" You can also bow yourself and use that as a signal to your dog.

Start by reviewing the touch stick with your dog in case she's a little rusty. Remember, she is supposed to touch the end of the stick with her nose. After a few repetitions, you are ready to move on to teaching bow. Gradually lower your stick each time she touches it, making sure to click and treat. Before long, the stick will be on the ground, which is just where you want it.

Many dogs will automatically go into the bow position with just that. If your dog is having trouble figuring out that she needs to drop her elbows, you can help her with a physical hint such as by putting slight pressure on her shoulders. This will encourage her to drop down in front. You want her elbows to be touching the ground.

If your dog lies down when she goes to touch the stick on the ground, just hold your free hand under her belly to keep her rear from coming down. (You can have the clicker in this hand if you need to.) You can also slip your foot under her belly before you try the trick again. Once she realizes she can't drop her rear, she'll raise it back to the bow position. Click and treat.

At this point, you can add your cue. After a while, your dog will figure out that when you give your cue, she is to drop her front end and keep her rear up. Once she is bowing consistently without the help of your hand, you can try it without the touch stick. She should bow on your cue. If she doesn't, repeat the exercise a few more times with the stick to make the trick stick in her mind.

Speak

Once again, have your clicker and treats ready before you begin. Just like many other tricks, speak comes naturally to some dogs. Of course, you'll have to settle for a bark as opposed to a conversation in English!

This trick can be helpful with dogs who are extremely yappy. Once you have taught them to bark on cue, you can also teach them to be quiet on cue. When your dog barks, you can tell her "good, now quiet!"

You may need to get your dog really excited before she will bark. Or you may have a dog who barks a lot. When she does bark, click and treat and say "good!" Repeat several times. It may take a while before she will bark without already being revved up, so be patient. Add your cue word only when your dog is consistently barking to get a treat. If you ask her to speak when she is calm and she doesn't bark, get her a little more excited and keep working at it. Eventually, when you tell her to speak, she should bark.

To make speak a little fancier, add a different cue word and put it in the form of a question. For example, you could tell her, "Bark once if you like parties." To teach this trick, simply use the cue "bark once" and click and treat only if she barks a single time. Then you can use the cue in all sorts of different situations. Since the cue is "bark once," you can change the questions, too. "Bark once if you want Dad's hamburger" or "Bark once if you like my new jeans."

Sit Up/Sit Pretty

Sit up (or "sit pretty") is the classic begging pose. Be aware that if your dog is older and has stiff joints, this trick is not a good choice. Also stay away from it if you have a long-backed dog such as a Dachshund or Basset Hound because these breeds run the risk of hurting themselves.

Have your clicker and treats ready. Now ask your dog to sit. Take a treat and hold it in front of her nose. Then move it up and back just a tad to get her to rock back onto her haunches. Once she is in position, you can click and treat or praise and treat.

Repeat several times using the treat to lure your dog into position. Then introduce your cue. Give the cue as you begin to lure her into position, then praise her for a job well done. Next, you can try luring your dog with your empty hand. After repeating a couple of times, you can gradually phase out your hand signal and have her work from your cue.

Cookie on the Nose

This trick is a bit more difficult because it requires your dog to hold still with a treat (we'll call it a cookie) on her nose until you release her. As you can imagine, this takes a lot of practice and self-control! For this trick, I recommend putting away the clicker. Dogs tend to get excited when they see a clicker, and overexcitement makes it harder for them to concentrate and be still.

Start by getting your dog used to you holding her head and nose. Pet her head, gently hold her muzzle in your hand, and give her lots of treats until she is comfortable with it. Now you can start teaching this trick.

Your first step is to teach your dog to balance the treat on her nose without moving. Choose a cue such as "wait" or "freeze"—but you won't use it just yet. Then, while you gently hold her muzzle steady, place a cookie on her nose. She may start to move, but try to keep her still. After a second or two of balancing the cookie, take it off her nose, praise her for being such a good dog, and give her a treat. If she moves before you want her to and knocks the treat off, don't let her get it. She needs to learn that she gets to eat the treat only when *you* say she can. Repeat several times.

As your dog gets the idea, begin to loosen your hold on her muzzle. If she is being good and staying still, you can even try moving your hand a few inches away. Don't expect her to stay still for very long, though—remember that you've just started teaching this trick. Gradually require her to stay still for longer periods and move your hand farther away from her before taking the treat off and rewarding her. You can add your cue word when your dog is holding still for several seconds without you needing to steady her muzzle.

Once your dog will stay still with the cookie on her nose for twenty or thirty seconds with you a few feet away, you can start giving her a release word, such as "okay!" and letting her eat the treat off her nose. It is important not to let her do this until she has learned to balance the treat, so that she will be less likely to cheat. If she moves and drops the treat before you release her, try to get to it before she does. You may need to stay close to your dog until she becomes reliable about waiting until you release her.

Some dogs will figure out how to flip the treat up and catch it, but others just aren't good catchers. It's a great trick either way!

5 Trickier Tricks

These tricks are, well, tricky. They require a bit more work and are more complex, so they're not for kids who don't love a challenge. However, this means they will be all the more impressive once you and your dog have mastered them!

Spin and Twirl

Spinning and twirling are simple in and of themselves, but they require your dog to understand the difference between turning one way and turning the other way. Basically, you want your dog to turn in a small circle in the direction you tell him to, on cue. I use "spin" for counterclockwise (in the opposite direction from the way the hands of a clock turn), and "twirl" for clockwise. Other possible cues are "twist" and "turn."

Start by teaching spin. Get some treats and your clicker and have your dog at your left side. With a treat in your left hand, lure his head so that he turns away from you, toward his left. When he has turned halfway, click and praise him for doing such a nice spin. Repeat several times. When you're sure he understands, use the treat to lure him all the way around, so he's made one complete turn. Repeat until he really gets it.

Now try giving a partial signal with your hand, not luring him the whole way around. If he spins all the way, click, treat, and have a big celebration. You can now gradually wean him off the lure.

When your dog is spinning reliably, you can add your verbal cue, such as "spin." Tell him to spin, and then, when he does, click and treat. Remember to reinforce only when he spins on your cue, not if he just spins at random. At first you may need to help him out with a flick of your left hand in the correct direction, but eventually he should be able to spin with just the verbal cue.

Next, get your dog to spin while facing you. Stand in front of your dog, signal with your left hand for him to turn counterclockwise, and give your verbal cue. He already knows the behavior and now is just learning it in a different setting. At first he may need help because he's facing a different way. But with a little practice, he will get it. Continue practicing until he can spin counterclockwise with, at most, a flick of your fingers pointing him the right way.

Now you can teach twirl. This time, have your dog on your right side when you begin. Lure his nose out away from you so that he is turning in a clockwise direction (the same way the hands move on the face of a clock). Click while he is turning. Repeat several times. When he is turning consistently, you can add your cue word. Then you can gradually wean him off the lure until he can twirl on his own.

As with the spin, step in front of your dog and help him twirl. As he gets more confident, you can once again wean him off the lure until the most you will need to get him going is a flick of your right hand in the correct direction.

Once he knows both spin and twirl, you can begin asking for either one when he is in front of you. At first he will be confused, but if he goes the wrong way, just help him do it right before you give him a treat. For example, if you say "spin" and he goes in the direction of "twirl," don't give him a treat. Be sure to reinforce him only when he does what you asked him to do. Also, don't always alternate which way you ask him to go—mix it up a bit. This will keep him paying attention!

You can also teach your dog to spin and twirl from a distance, which adds another layer of difficulty. Start out small by placing a target a foot or two away. Send him to the target, then tell him to spin or twirl. If he does it correctly, go to him and give him a treat. Increase the distance gradually, and remember to move the target a bit closer if your dog is having trouble.

In the beginning, you will probably have to use some hand signals as well as your verbal cues. But as your dog grows more confident, he will be able to spin and twirl on just your voice cues.

For this trick, you are looking for your dog to cover his face with a paw when you ask him, "Are you bashful?" Some other variations are "Are you shy?" or even "Are you guilty?"

Start by getting your treats and clicker ready, and grab either a stick-on note or a small piece of blue painter's tape (not duct tape or anything super-sticky!). Place the stick-on note or tape on top of your dog's nose, making sure that none of his whiskers get caught. When he swipes with his paw to get the note off, click and treat! Then put the note or tape back on his nose and try again.

This will require a lot of repetitions because the paw swipe is a reflex action. It will take a while before your dog begins to consciously think about what he is doing that causes you to click. Once he gets it, introduce your cue. Place the tape or note on his nose, give your cue, and click and treat when he swipes his nose with a paw.

After a lot of practice, try giving him your cue without putting the stick-on note on his nose. If he swipes his paw in response, have a big party! If not, go back a step and practice with the note or tape some more.

Once your dog is swiping his nose on cue, you can work on getting him to hold his paw there longer to give the effect of him hiding his face. When he puts his paw on his nose, click only if he holds it there for a second. When he is doing that consistently, require him to hold his paw on his nose for three seconds before you click and treat. Then you can ask for five seconds. Your dog can now cover his face for several seconds on cue!

Back

Back, meaning "walk backward," is a fun trick by itself. You can also put it together with other tricks. However, it does require your dog to have a good sense of balance and to know where his back feet are. You'd be surprised how many dogs don't! You'll need your clicker and treats for this one.

Start with your dog standing facing you, toe to toe. You will probably need to help him back up at first, because dogs don't usually do this on their own. Just take a little step forward. Don't step on his toes! Then, when he takes a step backward, click and treat. Repeat several times. When he is readily backing up when you step forward, you can start saying "back" each time before you step. After a while, your dog will get the idea and take a step back on his own when you start to move toward him. If he doesn't, add a slight lag between when you ask him to back up and when you step forward. Eventually, you want him to be able to back up on his own without needing the physical cue of you stepping forward. With some practice, he will begin to step backward before you step forward.

When your dog is consistently backing up on your cue, you can make the trick a little harder. Require him to take a few steps back before you click. Remember to increase the number of steps gradually—two, then three, and so on. If he gets confused, make it easier and require fewer steps for a few more tries. Eventually, he may be able to back up ten or fifteen feet without you taking a step!

If your dog has trouble backing up in a straight line, you can create a chute using furniture, or you can work along a wall, with the wall on the side he drifts to. A little practice like this will get him used to backing up straight, and his muscles will remember. Even so, you may need to go back to the chute or wall every now and then to refresh the idea of backing up in a straight line.

When your dog is good at backing up, you can try adding other tricks to make a little routine. For example, have him back up several steps, then tell him to spin or roll over. Another cool thing to try is getting him to back up while you back up in the opposite direction so you end up farther and farther apart. Also try having him back up with you while he's at your side.

Hold

Hold is a handy trick for your dog to know. You can use it for many different objects and situations, from holding a spatula to pose next to the barbecue grill to holding a pencil while you find some paper.

Start with a simple object that's easy for your dog to hold in his mouth, like a stick or a big wooden spoon. Hold it horizontally in front of you. When he touches it with his nose, click and treat. Repeat several times.

When your dog can do this reliably, wait until he puts his mouth on the stick. You may have to wait a bit, but eventually he will get impatient and grab it. When he does, click and treat. Repeat this several times.

Now you can ask him to take the stick in his mouth. He should catch on fairly quickly. When he is taking the stick in his mouth regularly, you can add your cue. Tell him "take it" before you offer the stick each time, and don't forget to reward him when he does it right!

Once your dog is taking the stick consistently on your cue, you can start asking him to hold on to it. This is easier to do without the clicker because the clicker gets many dogs excited and they open their mouths. Have him take the stick. When he has it, gently tilt his chin up with your hand and say "hold." Wait just an instant, then say "give" or "drop it" and give him his treat. Repeat several times. Work on getting him to hold the stick without your hand being there as a reminder.

Gradually increase the amount of time you ask your dog to hold the stick. Remember to reward him only if he lets go of the stick when you tell him to. You may need to occasionally remind him to hold the stick by tilting his chin up again as you say "hold." When he has a reliable hold with you close by and standing still, you can try walking around him. Be sure that whatever you ask him to do is fair. For example, don't walk out of the room and expect him to still be holding the stick a few minutes later.

At this point, you can begin introducing the hold with other objects.

Fetch

Fetch is a fun trick that can be used for many different things. You can teach your dog to fetch the newspaper, a person, specific objects—pretty much anything goes!

If you want your dog to fetch the newspaper, you need to get him used to holding it. Get together a rolled-up newspaper and your clicker and treats. Start by using the hold trick to teach him to hold the newspaper. Do it several times until you are sure he's okay with holding the newspaper.

Then put the newspaper on the ground and say "take it." At first, click even if your dog just touches it with his nose—it's harder to pick something up off the ground than it is to take it from a person's hand. Gradually increase what he has to do to get a click until he is picking the newspaper up off the ground. Once he is consistently picking up the paper, add your cue—something like "get the paper."

Now comes the fetch part. Be sure your dog will pick up the newspaper when you say "get the paper." Next, take a step or two back and put the paper down. Send him to the newspaper with the same cue. If he gets it and brings it to you, click and treat and have a party. Repeat several times. Gradually increase the distance you and your dog are from the newspaper when you send him.

If your dog doesn't want to bring the paper back to you, try upgrading from your usual training treats to better treats, like leftover steak. If the reward is good enough, he'll come to do a trade.

The key to getting your dog to fetch different objects is to name each object for him. That means each object has its own slightly different cue. My dogs all know the word "ball," so when I tell them "get your ball," they run off to find a ball and bring it to me.

To teach your dog the name of something, say the name of the object each time you have him take it in his mouth. He will figure out that the purpose of the game is to match the name with the object, and that when he does this, he gets a reward. For example, when you ask him to "get my slippers," you don't want him to run and grab the newspaper. If there is a mix-up, don't reward him and try again. Start closer to the item you wanted so your dog is less likely to be confused.

Fetching people is a fun trick. Dogs often learn our names anyway just from hearing us use them when talking with one another. To make sure your dog knows a person's name, have that person join you in a training session. Say the person's name—Fred, for example—and when your dog sniffs at Fred, click and treat. After a few repetitions, your dog will know to go to Fred when he hears you say "Fred." Then he will naturally come back to you for his treat.

You can use this trick to find a person or just to keep your dog busy. For example, during the day my mom often tells my young dog, "Go find Kate." My dog then spends several minutes searching the house for me, and she gets quite upset if she can't find me. Sometimes when I get home, she launches herself at me with more than her usual excited greeting. Then I know that she has been searching for me and is thrilled that she found me.

Leg Weave

This is an extra tricky trick. You want to teach your dog to weave back and forth through your legs as you walk. Obviously you won't be walking very fast! Also, this trick isn't recommended for large dogs, although if you think you and your dog can manage it safely, go for it.

Start with your dog on your left side. Take a big step forward with your right foot, and use a treat to lure your dog through the space between your legs. It is best to start by luring without a clicker until your dog gets the idea of what you are asking him to do. Dogs tend to get excited when working with the clicker, and leaping while trying to weave could result in you getting knocked over!

Your dog should now be on your right side. Take a step forward with your left foot, and once again lure him through the space between your legs. Give him a treat when he's partway through to let him know he's on the right track.

After several repetitions, introduce the cue you plan to use for this trick. Some possible cues are "weave" or "under." Step forward, give your cue, and lure your dog through your legs.

Now you can start using the clicker. With your dog at your left side, step forward with your right leg and give him your cue. If he ducks under your leg, click and treat when his head is right next to your right leg. If he doesn't make it, drop your right hand down with a treat so he can see it through the space. When he follows through, click and treat. Then step forward with your left leg and do the same thing.

Once your dog is weaving consistently with one click and treat per step and without your hand as a lure, start requiring him to go through your legs twice before you click. From there, you can gradually increase until you can walk at a somewhat steady pace with him weaving through your legs.

One way to make this trick easier with big dogs is to lift the foot you are stepping forward with off the ground so that there is a bigger space for him to go through. You may have to work on *your* balance for that!

The Alphabet

You probably can't teach your dog to read, but you *can* teach him to recognize the letters of the alphabet. Get your clicker and treats!

Start by writing a big letter A on a piece of paper. If your dog tends to get excited when he's working or likes to rip paper, you may want to paint the letter onto a piece of wood or something else more solid so that he won't shred it. Place your letter on the floor near your dog and wait until he sniffs or noses it, then click and treat. Repeat several times. Once he is going to the letter consistently, add your cue word, "A." Repeat several times until he is consistently nosing the letter when you give him your cue. Then put the A away.

Now write or paint a big letter B on a piece of paper or wood. Do the same thing you did with the letter A, only this time tell your dog "B." Don't forget to reinforce him when he touches it. All he has to do is touch it with his nose. Repeat several times.

Now comes the tricky part. Get out both letters and place them a little bit apart on the floor near your dog. First ask him to show you A. If he goes to the correct letter, click and treat. If not, wait patiently until he does. Repeat several times, always asking for A. Move the letters around occasionally so that he has to look at the letter and recognize it, and not just go back to the same spot each time.

It may take a while before your dog always goes to the letter A. Once he does, you can do the same with B. Then you can try alternating which letter you ask for.

Once your dog really knows the first two letters, you can gradually add more. Remember, this is going to take a while! However, eventually he will be able to recognize and pick out several letters (don't expect the whole alphabet). Then you can have him do tricks like spell out his name for an admiring crowd.

Find the Treat

For this trick, you need three plastic or paper cups and several dog treats. You want your dog to sniff each of the three cups and show you which one has the treat in it.

Start by placing one cup upside down on the floor in front of your dog. Take a good-smelling treat (such as cheese or meat—a plain biscuit probably isn't going to do the job) and put it underneath the cup. When your dog sniffs at the cup, make a big deal over it and lift it up to let him get the treat. Repeat several times.

Next, add a second cup without a treat under it. Make sure you know which cup has the treat! Allow your dog to sniff at both cups. When he settles on the one with the treat, praise him and give him the treat.

Repeat several times, then change around the order of the cups to make sure your dog is actually sniffing for the treat, not just going back to the same cup every time. Once he is doing that, you can add a third cup.

This is the easiest trick in the world for dogs to learn because they can smell the treat. Once your dog gets the idea, you can put a treat under the cup when he's not looking, and he'll still pick the right cup every time. The nose knows!

6 Spectacular Skits

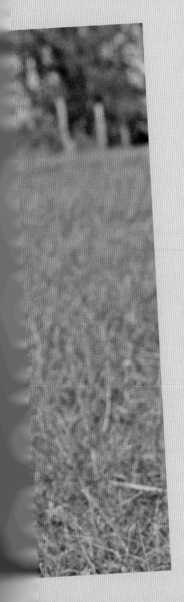

Once you and your dog know several tricks, you can put them together in interesting ways to make skits and scenes. Add some props and tell a little story, and pretty soon you've made up a whole play in which you and your dog are the stars.

Shut the Door

This trick can be very useful if you are carrying a bunch of stuff and your arms are full. Plus it's kind of cool. As well as the ever-present clicker and treats, you will need your target.

Start by reviewing the target with your dog. Just place it down, give her your target cue, and reward her when she goes to it a few times. Once her memory has been refreshed, you can move on.

Find a door that swings fairly easily on its hinges (it isn't fair to start your dog on a sticky door). Position yourself so that the door closes by pushing it away from you. You may also want to choose a door that isn't used regularly by the rest of your family so you and your dog won't be interrupted.

Since your dog already knows how to use the target, tape it to the closed door at about her head height. Close the door and then give her your cue for the target. When she touches the target, click and treat. Repeat several times.

When your dog is touching the target consistently, you can open the door a few inches. At first require her to move the door only a little when she touches it. As soon as it moves, click and treat. It may take a while before your dog gets excited enough to shove the door, so be patient. Once she is regularly moving the door when she touches it, you can start reinforcing her only if she pushes the door hard enough for it to close.

At this point, you can change your cue word to something that specifically means "shut the door." Something like "close" is best because it's easy to remember. From there you can gradually increase the amount you open the door until your dog can close it when it is open all the way.

Jump through a Hoop

For this trick, you'll need a Hula-Hoop as well as your normal supplies. Start by holding the Hula-Hoop upright on the ground—like a giant letter O with the bottom resting on the floor. Use a treat to lure your dog through it. Do this a couple of times, reinforcing her each time she walks through. Depending on the size of your dog, she may need to duck!

When your dog is walking through consistently, choose a word for your cue, such as "hoop" or "through." Give your cue, then wait for her to walk through. After she's done it a couple of times, you can switch to the clicker. Now just give her your cue and wait until she walks through the hoop. Then click and treat.

When your dog is consistently walking through the hoop on your cue, you can raise it off the ground just a few inches. Ask her to go through. This time she will have to step over the hoop. Repeat several times until you're sure she's comfortable. From there you can gradually raise the height a little bit at a time. Stop and practice at every change in height.

As the hoop gets higher, most dogs naturally switch from stepping through to jumping through. If your dog continues to step through the hoop instead of jumping, get her attention with a small squeaky toy or ball and toss it through the hoop. The dog will usually leap right after it.

About twelve to eighteen inches is the highest you should hold the hoop, and less for a small dog or an older dog who may have arthritis. If your dog is a puppy, don't raise the hoop more than few inches off the ground until she is a year old. Jumping is not good for young dogs. It can damage their growing bones.

Once your dog is a pro at jumping through the hoop, jazz up this trick a little. Red, orange, and yellow crepe paper taped or tied around the edges looks like fire, making your ordinary Hula-Hoop into a ring of fire. Be aware that when you decorate the hoop, it will look different to your dog. At first she may hesitate to jump through it, so you may need to lower the hoop and help her some. Before long, she'll be leaping through again.

Doggy Basketball

For doggy basketball, you need a basket and a ball. The baskets made for little kids are about the right height. You can probably find one at a garage sale. For a ball anything will do, from a tennis ball to a mini-basketball. Your dog should be able to hold on to it easily, but it should not be so small that she can accidentally swallow it.

First, get your dog familiar with the basket—especially the rim. Click and treat if she sniffs at or touches any part of the basket. When she's doing that consistently, wait until she touches just the rim before you click and treat. Repeat that several times. This may take a few days of practice.

Next, teach your dog to put her head into the basket, or "dunk." Initially you will need to click if her nose drops even a little way inside the rim. Gradually work her up to putting her whole head through. If your dog has a large head, just the nose will do—you wouldn't want her to get stuck! Add a verbal cue, such as "dunk" or "slam dunk."

Never, ever push your dog's head into the basket. If she never wants to do this trick, let it go and pick another one. No matter what, your dog is your best friend. No trick is worth hurting or scaring your dog.

Once your dog is doing the motions of the dunk, you can add the ball to the mix. First, teach her to hold and fetch the ball without the basket nearby (go back to chapter 5 for fetch). If your dog either doesn't like balls or goes nuts over them, this may take a while. Be patient, and don't try to move on until she has mastered the hold and fetch with the ball. Taking the ball and flinging it at you doesn't count!

Next, bring out the basket. This is where it gets tricky. Give your dog the ball, tell her to hold it, and then give her your "dunk" cue. She probably won't put all the pieces together at first, because this is a complicated trick. But it's worth a try.

If your dog doesn't get it, stand near the basket and have her bring the ball to you. Repeat a couple times. Next, have your hand right over the basket when she brings the ball to you. Repeat several times.

Your next step is to hold your hand beneath the basket so your dog drops the ball through the basket and into your hand. Give your "dunk" cue before she drops the ball each time, and click and treat when she does it. Repeat several times, and then gradually start moving your hand out from under the basket. Eventually she will drop the ball through the rim without your hand anywhere nearby.

Achoo!

Wouldn't it be cute if every time you sneezed, your dog ran over and brought you a tissue? This fun trick is really just a variation on fetch. Grab your clicker and treats, plus a tissue box. Get one of the square ones where you pull the tissue out through the top and the next one is always sticking up. Warning: You will probably go through a lot of tissues!

Start by placing the tissue box on the floor near your dog. When she sniffs at it or touches it with her nose, click and treat. Repeat several times until she is consistently touching the tissue box. The next step is to wait with your click until she touches the tissue that's sticking up—not just any part of the box. Repeat several times.

When your dog is doing that regularly, don't click until she grabs at the tissue. If she pulls it all the way out, make a big deal of it. As she gets quicker about grabbing the tissue, reinforce her only when she pulls it all the way out of the box. Repeat several times. Put each tissue that she pulls out either in the trash or to the side where your dog can't see it. This way she won't get confused about which tissue she is supposed to grab.

Many dogs find pulling the tissue out of the box really fun, much like when they pull the stuffing out of dog toys. Dogs especially enjoy doing tricks that they find fun. They are more exciting than the ones where the only appeal is that they get reinforced with a treat.

Encourage your dog to bring the tissue to you. As she gets better at pulling the tissue out, start clicking only if she brings it to you, not if she drops it right next to the box.

If your dog starts flinging tissues through the air, stop her! This is pretty funny, but it's not a behavior you want to encourage. Just say "no" in a firm voice and stop the game. Pick up the tissues and you're done for now. Your dog will figure out that if she throws the tissues all over the place, the game stops. Try again another time. It won't take long for your dog to realize that she needs to bring the tissue to you—especially if you have tasty treats to trade.

When your dog is bringing the tissue to you most of the time, add your cue. Your best choice is something like "achoo" or a fake sneeze. Slowly work up to giving the cue a little farther away from the tissue box.

At this point, reinforce your dog only when she has been signaled to get the tissue. Tissues you didn't ask for don't count, and you don't need twenty tissues for one sneeze! In the end, whenever you give your fake sneeze, she should run to grab a tissue for you. She may even do it after real sneezes!

Jump

Watching a dog gracefully jump is a beautiful thing, and most dogs seem to love it. Dog jumps can be made out of almost anything, from a broomstick set on two buckets to a hay bale. There are all sorts of fun things your dog can jump over. You can even build a special jump.

Before you get started, there are a few things to remember. First, don't encourage your dog to jump over fences because you will probably regret it. When teaching your dog to jump, start with the jump very low and work up slowly. You wouldn't want her to get hurt. Don't ask young dogs to jump higher than a few inches until they are at least 1 year old. In fact, it's best to wait until they are 1½ years old before going to full height.

How high is full height? It depends on your dog. Your best bet is to ask your dog to jump something that is only as high as she is tall—in other words, nothing higher than the top of her shoulders. Also consider your dog's shape. Dogs with long backs and short legs, such as Dachshunds, Basset Hounds, and Corgis, are more likely to have trouble jumping. Overweight and older dogs also should not be asked to jump very high. Others may not want to jump at all. Respect your dog's wishes.

Start by creating a jump—a pole or a broomstick is fine—and set it on the ground. Lure your dog over the jump and reinforce her with a treat. When she's done this a few times, introduce your jump cue—"over" and "jump" are two possibilities.

After a few repetitions, you can try giving your dog just the cue, and then wait until she goes over. When she does, click and treat. Repeat several times. As she gets more confident, you can slowly increase the height of the jump. Do this very gradually—just a few inches at a time. With an adult dog who is in good condition, you can probably add about two inches a week.

If your dog looks at a jump and refuses to go over because it is too high, lower the jump height and work on that for a while until you are sure she is ready to go on. Most dogs refuse to jump not because they are lazy, but because they aren't sure they can jump that height without hurting themselves. Never force your dog to jump over something she doesn't think she can jump.

Once your dog knows how to jump on cue, you can work on a jumping routine. Put several jumps in a row, or try figure-8s and circles. Just be sure to give her enough space to land safely between jumps.

Jumping over and through things is part of a canine sport called agility. Agility dogs also go through tunnels, weave between poles, and walk on seesaws and other equipment. If you have seen agility on television or in real life and think it's something you and your dog would like to get involved in, look up a local dog club and see about joining a beginner class. This way you can learn from an experienced instructor who can work with you and your dog on equipment that is safe and stable. Don't try to reproduce what you see on TV unless you are absolutely sure that what you are asking your dog to do is safe and fair.

Dancing with Dogs

When your dog has learned more than one trick, you can start putting several tricks together. There's even a dog sport where people do this—to music. It's called musical freestyle, and people and dogs "dance" to music. Basically, they do a series of tricks and behaviors in time to a song. What's really fun about it is that the dogs really seem to get into it. Some dogs even have favorite songs. Musical freestyle can be really fun to fool around with, even if you're just listening to the radio and doing various tricks, such as spins and twirls, that seem to fit with the music.

Start by putting on a favorite song and practice heeling (walking with your dog on your left side) to the beat of the music. From there, throw in a few spins and twirls at your side to get a feel for how much time it takes for your dog to perform her tricks.

For either freestyle or just performing for your friends and family, you may want a routine. That's a fixed plan of what you and your dog will do and when. When you come up with a routine, go through each step in the right order with your dog, and reinforce her after each trick. Remember, although you know all the tricks will be put together into one big routine, she doesn't yet.

Go through the whole routine, doing each trick individually a few times to make sure your dog remembers them all. When she does, you can begin to go through the routine a little faster, stopping to give her a treat after every few tricks. Even though you won't be reinforcing her after every trick, your dog will stay interested as long as you praise her. Not knowing when she is going to get the next treat will also keep her attention on you because she won't want to miss a treat. You can gradually increase the number of tricks she does between treats until she can do the whole routine in one go. Remember to praise her (maybe say "good dog!") after each trick, even though she doesn't get a treat until the end.

After you practice your whole routine a few times, you will probably find that your dog remembers the order of the tricks. Dogs like this sort of game, and yours may start going from trick to trick even without a signal from you because she knows what's next. To keep her happy, every now and then give her a treat in the middle of your routine. It will come as a pleasant surprise.

Some tricks blend together very smoothly. For example, backing up is a good setup for many other tricks. You could have your dog back up and then spin, twirl, roll over, wave, bow, or sit up—all at whatever distance your dog backed up to. The first couple of times you try this, you may need to remind your dog to stay away and not creep back toward you.

Spins and twirls flow together fairly well, too, creating a picture of a canine tornado. Several spins in a row look like your dog is chasing her tail. I like to have one of my dogs do shake and wave one right after the other, without me actually taking her paw for a shake. If your signals are timed right, your dog will look like she is dancing in time with the music.

You could leave your dog on a stay, call her to you, have her sit, and then shake her paw for a job well done. There are many combinations, and anything goes. Most dogs enjoy working with their people, and for them it's just playing. When you can work together happily, it opens the door to many opportunities for showing off your dog's skills and just having fun together.

Fantastic Felines

When people think of pet tricks, they usually think of dogs. But cats can do tricks, too, and are often very good at them. Although they may be independent, cats are very smart and know a lot more than they let on. However, cats are different from dogs in many ways, and you need to understand what those differences are.

For one thing, cats tend to be pickier eaters. Bits of dry cat food will do for training treats if nothing else is available and the cat hasn't eaten yet that day. Otherwise, the treat had better be something *really* good. Pet supply stores sell various types of kitty treats that you can offer your cat to find which he likes best. Small pieces of deli meat often go over well, too.

You will need to keep giving your cat food rewards even after he has learned a trick. Cats don't work for free!

Some cats really like toys, and you can work them into your training program. Instead of doing a click and treat, you would click and throw a jingle ball or a toy mouse. Feathers go over well. Many cats enjoy a ball of crumpled-up paper or aluminum foil.

For the most part, your cat can do anything a dog can do. Barking on cue may be hard to teach, but a lot of other things that are traditionally considered "dog" tricks can also be taught to cats. I've described some tricks here, but look over the tricks in chapters 4, 5, and 6 and think about which ones would work well for your cat.

Clicker training works well with cats. The touch stick is also very useful, although your cat may prefer to touch or bat it with his paw instead of touching it with his nose. After all, that just wouldn't be dignified! Before getting into the tricks, introduce your cat to the clicker and touch stick the way I described in chapters 2 and 3.

Shake

Yup, cats can shake paws, too. Get together some treats or your cat's favorite toy and your clicker. Now you're ready to get started.

The first step is to get your cat to lift his paw off the ground even a little bit. You may have to wait a while, depending on how soon your cat decides he is bored and moves. As soon as one of his front paws moves, click and treat. Then you should have his attention. Repeat several times until he is lifting his paw again and again, looking for a treat. What you want is for him to decide to lift his paw on purpose—not just do it because he is on his way somewhere else.

When your cat is lifting his paw consistently, put your hand out right next to his paw. When he moves his paw, slide your hand under it. Click and treat or throw a toy for him. Repeat this a couple of times. After a few repetitions, don't move your hand to meet his paw. Instead, wait and see if he will put his paw in your hand. If he doesn't do it right away, you may need to repeat the step where you put your hand under his paw. Do this a few more times. When he puts his paw in your hand, click and offer a reward.

Once your cat is putting his paw in your hand regularly, gradually raise your hand so that he has to lift his paw higher to put it in your hand and get a reward. Remember to lower your hand a little if he seems to be having trouble.

At this point, you can add your cue, or you can just use the hand signal, holding your hand out to him, palm up. Most cats need a signal or a physical cue as well as a verbal cue, although a few will work for verbal cues only. (That's another way in which cats are different from dogs.)

High-Five

You can teach your cat high-five the same way you teach shake. To get him to give an actual high-five as opposed to a "low-five," you can gradually turn your hand so that your palm is facing him. Remember to always click and treat when he does it right! After you have repeated this several times, he will see the position of your hand and know that it's time to give you a high-five.

Spin and Twirl

For these two tricks, you will need something to lure your cat. You can try a touch stick or a cat toy with a feather on the end of a stick. Start by playing with your cat, dragging the toy across the floor and getting him to chase it. When he's following the toy, use it to lure him in a small circle. He should follow it. Then play some more.

You can gradually make the circle smaller until your cat is spinning as if he is chasing his tail. In fact, he might get distracted by his tail! Be sure to have him go both directions. This may turn out to be his favorite game.

You can use a laser pointer to teach a cat to spin and twirl. Some cats go crazy for the laser dot and will happily chase it. There are just two things to be careful about. The first is never shine the laser into the cat's eyes. The second is that since there's nothing for the cat to catch when he's chasing a laser, end each training session by having the laser dot land on a toy that your cat can pounce on. If your cat becomes too revved up by the laser to relax when the training session is over, switch to another toy or a favorite treat as a training lure.

Bow

For bow, you will need your touch stick and a clicker. To make the stick more interesting, you may want to attach a feather or toy of some sort to the end, or just use a feather wand toy.

Start by playing with your cat a little. When he pounces on the end of the touch stick, he may automatically go into the bow position. If that is the case, click! Then let him play some more or give him a treat. Repeat several times.

If your cat doesn't do the bow on his own, try bringing the stick suddenly back toward him. This should make him go into the bow position. Click and treat or play some more. Repeat several times to reinforce that this is what you want him to do.

Leg Weave

This is another trick that is easy to teach your cat by using a lure. Cats tend to weave in and out of our legs anyway—especially when we're making their supper.

Find something your cat likes and get him playing with it. Have your cat on your left side and a toy on a stick or string in your right hand. This will be your lure. Take a step forward with your right leg, and wave the toy so that he jumps through the space between your legs. He should now be on your right side. Take a step forward with your left foot and transfer the toy to your left hand, once again dangling it so that he follows between your legs.

You can also do this standing still. Use the lure to have your cat weave a figure-8 around your feet. It's probably best to wear pants when you're teaching this trick, in case he misses and grabs your leg instead of the toy. If he does, remember it was an accident and don't be angry. Just try to keep the toy away from your body.

Sit Up/Sit Pretty

Using either a treat or a toy, lure your cat up and back so that he is perched on his back legs, then click and treat. Repeat several times. Make a sweeping motion as you lure him up so that later you can use this motion as a signal that you want him to sit up even if you don't have the toy with you. Just remember, cats usually require a reward every time they do a trick!

Jump

Cats are born jumpers, and they love to jump. You teach a cat to jump the same way you teach a dog. Go back to page 62 and you'll see how. Review the jump through a hoop trick, too (page 57), because cats certainly can jump through hoops. The big difference is that you may not be able to put the jump on a cue. Use a toy to lure your cat over the jumps. Also take breaks now and then to let your cat catch the toy and play a little—the chase is no fun if he never gets to catch it!

Use a toy to lure the cat over the jumps. Most cats are able to catch the toy often enough to stay highly interested. If your cat isn't able to catch the toy often, you might want to take frequent breaks to let your cat catch the toy and play a little. If your cat is overweight or out of shape, give him breaks to rest.

When working with jumps, start low and gradually increase the height if your cat seems comfortable with it. Cats can jump a lot higher than dogs. They can also jump from one raised place to another—say, from the couch to a chair. Remember not to ask kittens to jump over anything very high. Wait until the cat is at least 1 year old before you start high jumps.

Cats also do agility as a sport. With a touch stick or a toy on a string, you can lure your cat over jumps and ramps and through tunnels. On a rainy day, set up a cat agility course indoors and see how much your cat enjoys it!

Down on the Farm

When the animal is more confident about taking treats from your hand, begin to talk softly. What you say doesn't matter; she just needs to get used to your voice and learn that you are not scary, even when you're talking. Then you can gradually inch your hand closer to your body so that she has to come closer to you to get the treats.

When she is willing to come right up to you for treats, try slowly reaching your other hand toward her while she is eating. If she backs away, freeze your hand where it is and continue to wait. The next couple of times she comes up to your hand with the food, don't move your other hand. When she no longer seems concerned about the other hand, you can try moving it closer again.

Eventually, you'll be able to hold your free hand in a position where the animal has to brush against it to eat. This small amount of contact is enough to teach most tricks. But it's even better if you can gradually get her used to being touched and handled. Be aware that it could take a couple of weeks before a shy animal is totally comfortable around you. Just be patient.

You can use all sorts of food for treats. For a horse, you can use horse treats, grain, carrot pieces, or apple slices. Cheerios work well for sheep and goats. Chopped up bits of fruit or vegetables go over well with many animals. For ducks and chickens, bright-colored tidbits such as dried corn, fresh or frozen peas, shelled sunflower seeds, berries, or almost any kind of grain work well. Try different treats until you find something your animal enjoys.

Start by teaching the animal what a clicker means, the way I described in chapter 2. Remember that most farm animals are prey animals—the kind of animals that other animals eat—which makes them naturally cautious. So getting your farm friend used to the clicker may take longer than it would with a dog or cat. Try using a quieter clicker, or mute the sound by clicking it inside your pocket or a cloth bag to make the sound less scary.

Shake

Horses, mules, donkeys, miniature horses, goats, cows, and sheep can all learn to shake. Get your clicker and treats ready.

Wait until your animal moves her hoof. When she does, click and treat. Repeat several times until she is moving her foot on purpose to get a treat. Then start clicking only if she lifts her hoof all the way off the ground. Once she is doing that consistently, require her to hold her hoof up for a few seconds before you click and reward her.

At this point, you can introduce your cue. Give your cue, wait until she raises her hoof and holds it out for a few seconds, and then click and treat. Repeat a few times. Now, when she holds her hoof out, take it in your hand and click and treat. Be sure to click and, if possible, give her the treat while you are holding her hoof to let her know that this is what you want. Repeat several times. After that, reinforce her only if she raises her hoof when you ask her to.

Bow

Bow works for horses, mules, donkeys, miniature horses, goats, cows, and sheep. For this trick, you will need a touch stick. Start by teaching your animal to touch the touch stick with her nose when you say "touch." (I described how to do this in chapter 3.) At first, you will hold the stick out in front of her, about as high as her head, for her to touch it.

Once she understands how you are using the touch stick, begin to gradually lower the stick. Your goal is for her to touch the end of the stick when it is held down near her front hooves. Be patient and go slowly. You might end up lowering the stick just a few inches each day.

When she is consistently touching the stick down by the ground, switch your cue from "touch" for the touch stick to "bow." Repeat several times, and don't forget to click and treat. Next, put the touch stick away and try just giving the cue. If she doesn't bow, get the stick back out and work on that some more before trying again without the stick.

Where's My Hanky?

The tools you'll need for this trick are a piece of cotton cloth that resembles a hanky and some tasty treats like cut-up pieces of apple or carrot. Use a sturdy piece of cloth that won't rip or tear easily, as some horses may bite down on the cloth quite hard the first time it's presented to them.

It's best to teach your horse this trick in his stall or while he's in a halter and lead shank so he's more likely to stay facing you.

Follow these steps to teach the trick:

1. Place the treat in the cloth so that it sticks out of the top of the cloth. Present the cloth with the treat to your horse as you ask, "Where's my hanky?" When he takes the cloth with the treat, praise him. Give him another tasty treat as you retrieve the cloth.
2. Repeat until he's quickly taking the hanky with the treat.
3. Now, present just the hanky. When he grabs it, be sure to respond quickly with a treat and praise.
4. To jazz up the trick, you can slowly transfer the hanky from your hand to a pocket or to a nearby shelf or other area. Each session, move your hand with the hanky either closer to your pocket or closer to the shelf. Some horses will quickly learn to take the hanky from wherever you place it, and others will need very small transitions from one placement to the next. Be patient. Always ask, "Where's my hanky?" to cue your horse.

Once your horse learns the behavior, you can put together a funny skit. Imagine a friend is with you in the barn. You fake a sneeze, look toward your horse, and ask him, "Where's my hanky?" and he retrieves it from the shelf for you!

It's easy to transfer the behavior of picking up a hanky to picking up other objects. Take a brush and ask the question, "What would you like?" to get your horse to pick up a brush for you to groom him. Or use a small flag for him to wave on the Fourth of July. Simply start at step 1 with the new item and the new cue. One word of caution, though: Be sure to reward the behavior only when you give the verbal cue. Otherwise, you may find your horse picking up objects without being asked!

Come

This one is for ducks and chickens, although other critters can learn it as well. These fine-feathered fowl love food, and you can use this to your advantage.

Start your training session at mealtime. To teach a group of fowl, scatter food on the ground or pour it into their food dish. When they come to you when they see you with food, add whatever cue you would like to use. Maybe "Here, duck, duck, duck" or "Dinner!" After a couple of days, they will come running when you call. Always have something tasty to give them!

You can also teach an individual chicken or duck to come. Your first step is to get her alone and in a small, confined space, such as a stall or a pen. Have some seeds, berries, or whatever she likes to eat. If she is afraid of you, start by getting her used to your presence the way I described at the beginning of this chapter. Work on getting her to come up to you. Once she's got that down, you can start working on come.

Hold your hand out flat with a treat on it. When your bird comes to you, give her the treat. Repeat several times. Now add the cue you have chosen. When she is consistently coming when called, try going to a larger enclosure. Stop by at various times during the day to call to her, always with treats ready. Before long, she'll come reliably no matter where she is.

Weird and Wacky

There are many other kinds of animal companions, and you can teach tricks to all of them. Find treats that your animal friend likes and spend some time getting him familiar with the clicker, the way I described at the beginning of chapter 2. Now you're ready to go!

Turtle High-Five

Get some treats that your turtle loves. Pick a time of day for training when he is active and hungry. Go to a spot where he is comfortable and relaxed. If his head and/or legs are pulled into his shell, he probably isn't happy!

Remember that you are going to have to be patient with your turtle. Make sure you have plenty of time to train before starting. Start with a quick review of the clicker. When he hears a click, he should look to your hand for a treat. Some turtles like to work for touch, and if that's what yours likes, just pet or stroke him instead of giving a treat.

Now place your free hand (the one that's not holding the treat) a few inches away, with your palm facing toward your turtle. Wait for him to move a front foot to step closer to your hand. When he does, click and treat. You don't need a big motion just yet—lifting his foot just a little will do. Repeat until he is consistently raising his foot to come to you to get a treat.

The next step is to click only if he touches your hand with his foot. Be patient! Also be sure that your hand is close enough for him to reach it. Repeat until he is reliably hitting your hand with his foot.

Remember to thoroughly wash your hands whenever you handle your turtle, because reptiles can carry germs that may make people very sick. And please, never pick up a wild turtle and try to keep him as a pet. It's cruel to the turtle—and illegal!

Turtle Soccer

For this trick, you will need your clicker, treats that your turtle likes, and a lightweight ball that will be easy for him to move. Ping-Pong balls are about perfect for most small turtles. They seem to be attracted to the white color—maybe the Ping-Pong ball looks like a turtle egg to them. As always, practice during a time of day when your turtle is active and hungry.

Start by placing the ball in front of your turtle. Wait for him to investigate it, and then click and treat. It doesn't matter if he touches the ball with his nose or his foot—there are no official rules for turtle soccer! Repeat several times.

The next step is to reinforce your turtle only if he hits the ball hard enough to move it. Repeat until he is consistently moving the ball when he touches it. Some turtles really get into this game and go nuts whacking the ball around. It can be a lot of fun!

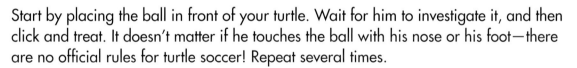

Fish Come

Though it may seem simple, you will be amazed at how many people are impressed by a fish who comes when called. Start by dividing the amount of food your fish gets each day into three portions. This gives you three opportunities to practice each day.

Place some sort of visual marker—an object your fish can easily see—at one end of the tank. A bright bead, a piece of coral, or a pirate ship would work. This marker will show your fish where to go when he is called to get his food. Be sure that your marker isn't something that will mess up the water chemistry of your fish's tank or could contain toxins. For example, don't use a rock that you found and rinsed off, because it still isn't as clean as your fish needs. And it may leak minerals into the water that are bad for your fish.

Next, take one portion of food and very lightly tap the surface of the water right above your marker while calling "Here, fishy fishy," or some other cue. Then drop the food. Your fish will come running—er, swimming—to get his meal. A little while later, repeat with the second portion. Later in the day, try again with the third. Repeat over several days. Before long, your fish will learn that a finger tapping the water means food will come to a specific spot in the tank.

Always give your cue before you tap. After several days, add a slight pause between the cue and the tap. Over time, you should be able to stop tapping and your fish will respond just to your voice. As long as you are patient, he will eventually come when you call.

Fish Agility

You can do an underwater version of dog agility for fish right in the tank. To start, you will need to teach your fish about the clicker and touch stick, as I described in chapters 2 and 3. Use a tiny bit of fish food as his reward each time. Practice at mealtimes when he is hungry, using his dinner for treats. You can click above water, and your fish will hear it. Use the touch stick to reach down underwater.

For equipment, you can make little hurdles, poles your fish can weave in and out of, and even a tunnel. Because your fish is underwater, he can swim under the hurdles instead of over them. You can also set up bars like football goalposts that he can swim though. Choose equipment that is sturdy and can't be knocked over, and big enough that your fish won't get stuck. It should be safe for a fish tank as well.

Weave poles are fairly easy for a fish to master once he is familiar with the touch stick. Set the poles up in his tank, a few inches apart. Then use the stick to lead him around the poles. At first you should click and treat after each pole, but then you can gradually increase the number of poles he has to do before getting a reward.

For swimming through goalposts, just lead him through with the touch stick. For swimming under hurdles, you will first need to slide the stick underneath while your fish isn't looking. Then, when he swims up to it, pull it underneath the hurdle so he will follow. Repeat several times. Then you can try putting the stick on the opposite side of the hurdle from your fish and reinforce him only if he swims underneath. Remember to be patient!

Teaching your fish to swim through a tunnel is just like teaching him to swim under a hurdle. Start by sliding the stick through when he isn't looking, and then lead him through the tunnel several times. With practice, you can get him to swim through just with you running the stick along the side of the tunnel. A clear tunnel works best for this.

Be creative, and see if you can come up with new obstacles to teach him.

Sing

Obviously, this trick is for birds. Start by teaching your bird that the clicker means treats are coming his way, as described in chapter 2. Once he has that down, you can move on.

The first step is to wait until your bird chirps or whistles. Depending on your bird, it may take a while. Wait patiently, and when he makes a sound, click and treat. Then wait again. Repeat until he is chirping regularly to get a treat.

At this point, you can add your cue. Something like "sing" or "chirp" is simple and easy to remember. Repeat several times. Give the cue first, and then click and treat when he sings.

Step Up

For this trick, you will need your clicker, treats, and some sort of perch for your bird, such as a stick or dowel. The perch should be a comfortable size for your bird to grab onto with his feet.

Hold the perch near where your bird is sitting, and—you guessed it!—wait. If he steps right onto the perch, click and treat. However, he probably won't. In this case, you will need to click and treat just if he looks at the perch and shows some interest. Repeat several times. Now you can require him to lift a foot before you click and treat. Repeat this a couple of times.

Now, click and treat only if your bird touches the perch with his foot. When he is doing that, the next step is to click if he grabs the perch with his foot. Once he is doing that consistently, wait until he is stepping onto the perch with both feet before you click and treat. Repeat until he reliably steps onto the perch when it's offered. Don't forget to click and treat!

Now you can introduce your cue word. Use something that will be easy to remember, like "step up." Give your cue, and when he steps onto the perch, click and treat. Only reward him for stepping up when you ask him to.

10 Take Your Show on the Road

nce you and your animal friend have learned an impressive array of tricks, you may want to perform them for people other than your family and friends. Be sure to think this decision through and talk it over with your parents. You must consider many different things before taking your show on the road.

Before you commit to a public performance, practice your tricks in a bunch of different places: different rooms of the house, out in the yard, at a friend's or relative's house, at the park. This will build your animal companion's confidence in new places. It's not so easy for her to do the same tricks she learned at home when she is in a different environment. Be sure to give her plenty of practice.

Health and Safety

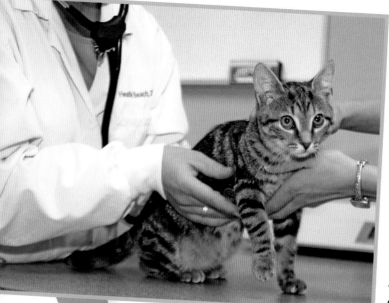

Before going anywhere with your pet, make sure that she is happy and healthy. A sick pet isn't going to do tricks, and as a responsible guardian, you should keep her at home where she will be happiest. Have your pet checked by your veterinarian, especially for parasites or diseases that could spread to people. For dogs and cats, you may want to apply a flea and tick preventive, depending on the area you're going to. Be sure to ask your vet about this.

If your pet has a health condition that requires medication or treatment at various times during the day, be sure to bring along anything you might need to give her. It may be better to keep her at home, though. She is just as good a trickster no matter where she performs.

Always have a collar ID on your cat or dog, but especially when you leave your home. Ask your veterinarian about getting your pet a microchip. This tiny device goes under the skin and contains all the information needed to bring her back to you if she gets lost.

Where You Can Go

- **Trick Contests.** Many pet supply stores and dog clubs have trick contests. They are usually a lot of fun and often give away great prizes. Keep a lookout for fliers or posters announcing them. For shy pets who don't like to go to new places, there are many online trick contests that use video recordings.

- **Demonstrations.** Contact your local dog club and talk to them about upcoming events where they are having trick dog demonstrations. The same goes for cat shows. And county fairs sometimes offer a trick division for farm animals. Another place to check for places to perform is your local animal shelter. Shelters often have open houses, walkathons, and other events where your participation would be welcomed.

- **Nursing Homes and Hospitals.** Nursing homes often have a Pet Day when people from the local community are allowed to bring their pets to visit. Hospital patients and nursing home residents love to see the animals, especially if they know cool tricks. Once again, be sure that your dog is people-friendly and is okay around things like walkers, oxygen tanks, and wheelchairs. You'll also need to get certified, which I will explain a little later in this chapter.

What to Bring

Always bring a few things with you when you and your pet go somewhere together to perform tricks.

- **Water.** Even if it isn't supposed to be hot out, being in an unfamiliar place can be stressful for your pet, and she may get hot from working. Bring plenty of water from home, as well as a bowl or dish that she can drink from.

- **Treats.** You will want treats while you warm your pet up and get her ready to perform. If she is nervous, you may need treats to keep her focused and happy.

- **Props.** These include your clicker, touch stick, and everything else you will need—and possibly a few extra things. It's kind of hard to show off the tissue trick if you don't have a box of tissues!

- **Vaccination records.** Depending on where you are performing, vaccination records may be required. No matter what, it is a good idea to keep a copy in the family car.

- **Crate or carrier.** You need a way to keep your pet safe if you have to do something without her. A crate or carrier also provides a nice, quiet hiding place for your pet when she isn't performing.

- **Cleaning supplies.** Bring along paper towels, poop bags, and a spray cleaner that takes care of pet odors. Animals in unfamiliar places sometimes have accidents—even pets who are completely reliable at home.

Certification

If you want to visit nursing homes or hospitals, your pet will probably need to be certified. Call the facility to find out what kind of certification you might need and whether they have any age restrictions for people visiting with therapy animals. Then contact one of the major groups that are involved with animal therapy work, such as the Delta Society or Therapy Dogs International. (You'll find their contact information in the Resources section of this book.)

You will need to fill out a lot of paperwork, and you and your pet will be tested to be sure you will be a good team. Before doing anything, be sure that your parents are comfortable with you and your pet doing this sort of work and are willing to provide transportation. You will probably need a parent to sign most of the forms, too. Though it may seem like a hassle, in the end it is more than worth it because these groups provide insurance for you and your pet. Plus, it will look great on college applications!

One Last Thing

When you and your pet go somewhere new, remember that this is a fresh and exciting experience for her. Expect that you will need to help her out a little, especially the first few times you go somewhere. She will need to get used to the hustle and bustle of being away from home.

And if a trick goes horribly wrong, just laugh. No matter what, your pet will always be one of your best buddies. If you make a mistake, she'll still talk to you when you go home. You owe her that, too. Forgive and forget, or save the story to tell over dinner someday and make everyone laugh. There's always tomorrow.

Resources

Books about Clicker Training

Alexander, Melissa, *Click for Joy*, Sunshine Books, 2003.

Book, Mandy, and Cheryl Smith, *Quick Clicks: 40 Fast and Fun Behaviors to Train with a Clicker*, Dogwise Publishing, 2001.

Kurland, Alexandra, *Clicker Training for Your Horse,* Sunshine Books, 2007.

Pryor, Karen, *Getting Started: Clicker Training for Cats,* Sunshine Books, 2003.

Pryor, Karen, *Getting Started: Clicker Training for Dogs,* Sunshine Books, 2005.

Books about Trick Training

Baer, Ted, *How to Teach Your Old Dog New Tricks*, Barrons, 1991.

Coile, Caroline, *Silly Dog Tricks: Fun for You and Your Best Friend,* Sterling Publishing, 2006.

Fields-Babineau, Miriam, *Cat Training in 10 Minutes*, TFH Publications, 2003.

Fletcher, Carole, *Trickonometry: The Secrets of Teaching Your Horse Tricks*, Horse Hollow Press, 2003.

Sundance, Kyra, *101 Dog Tricks: Step-by-Step Activities to Engage, Challenge, and Bond with Your Dog*, Quayside, 2007.

Web Sites

Delta Society
www.deltasociety.org
This organization works to improve people's quality of life by certifying service and therapy animals. It is a good group to get involved with if you would like to do therapy visits with your animal friend.

Karen's Corner
www.clickertraining.com/karen
Karen Pryor is one of the people who developed clicker training, so check out her web site for some extra info.

Therapy Dogs International
www.tdi-dog.org
TDI is similar to Delta Society but is for dogs only.

About the Authors

Kate Eldredge

Kate Eldredge is a high school student who has lived with dogs and an assortment of other animals all her life. She enjoys working with them and has had a great time teaching tricks to all sorts of animals, from her dogs to donkeys to a goat! She thinks tricks are a great way to have fun with your pet and wishes kids everywhere the best of luck teaching their pets.

Jacque Lynn Schultz, CPDT

Jacque Lynn Schultz, CPDT, has worked for the ASPCA for more than twenty years. She began as a behavior counselor, then headed the Training and Behavior Department, and is now a senior director in the Community Outreach Department. In her current role, she works with animal shelters across the country to enrich and improve the lives of the animals in their care. Jacque is a popular speaker, an award-winning author, and a certified pet dog trainer. She lives in Brooklyn with Bop, a sassy white cat with orange tabby patches, and Harpo, a blonde Tibetan Spaniel who is a master of many tricks.

About the ASPCA

The American Society for the Prevention of Cruelty to Animals (ASPCA) was founded in 1866 by a wealthy man from New York City named Henry Bergh. Henry first got the idea of devoting his life to protecting animals when he was serving as a diplomat in Russia. While riding in a fancy horse-drawn coach, Henry saw a peasant beating a lame cart horse on the side of the road. The horse was injured and couldn't put weight on one of his legs, but the peasant was beating him to force him to continue pulling the cart. Henry ordered the man to stop. Russian peasants did not dare argue with noblemen, so the man did as he was told. In that moment Henry Bergh felt great joy at being able to help a suffering animal.

Henry resigned his diplomatic post and returned to the United States where he worked tirelessly to set up a society to protect animals and to convince the New York State Legislature to pass a law making it a crime to beat and overwork animals. The legislature not only passed a law but gave the ASPCA the police power to enforce the new law throughout the state. The ASPCA was the first animal protection organization in the Western Hemisphere, and Henry Bergh was the first person to enforce the law on the streets.

More than 140 years later, the ASPCA is still enforcing the anticruelty law through its Humane Law Enforcement officers—sometimes called "animal cops"—who investigate complaints and arrest people who are hurting or neglecting their animals. But the ASPCA protects animals in many other ways, too. One of the most important is through its Humane Education department. Humane education means teaching children to care about animals in their own homes and in their communities. It fosters kindness, compassion, and respect for animals, the environment, and other people. Humane education tries to build a sense of responsibility in young people to make the world a better, kinder place.

Here are some exciting ASPCA Humane Education programs to check out.

The **ASPCA Henry Bergh Children's Book Award** (www.aspca.org/bookaward) was set up to honor new books for children and young adults that promote compassion and respect for all living things. Each year the ASPCA gives awards in six categories: Companion Animals, Humane Heroes, Ecology and the Environment, Poetry, Young Adult, and Illustration. It gives separate awards for fiction and nonfiction. Winning books bear the Book Award seal, which consists of a silhouette of a horse and a gentleman wearing a top hat. The awards are named for Henry Bergh because he not only founded the ASPCA to prevent cruelty to animals, but also helped found the New York Society for the Prevention of Cruelty to Children in 1874. It was the first society to protect children from being beaten and starved by their parents. Henry Bergh understood that laws and police powers were not enough to ensure a humane society. Children must be treated kindly and taught to be kind to all living things in turn.

Henry's Book Club (www.aspca.org/ henrysbookclub) was formed in 2008 as a way for kids who like animals *and* books to get together to read the ASPCA Henry Bergh Children's Book Award winners and discuss them as a group. Kids meet once a month in clubs that have been organized in schools or communities. The ASPCA provides polls, quizzes, and discussion questions for use at club meetings. Teens over 13 have the option of joining a virtual club through the ASPCA Online Community, where they can chat live about the books with teens across the country—and even with the authors, who join in from time to time as special guests!

ASPCA Kids, Animals, and Literature Bibliography (www.aspca.org/bibliography) is an online list of books about animals that the ASPCA Humane Education staff has reviewed and recommends as accurate, humane, and fun to read. It contains hundreds of titles and is easy to search by subject, title, author, or age, or by whether the book is fiction, nonfiction, or poetry. Winners of the ASPCA Henry Bergh Children's Book Awards are listed, too.

ASPCA Animaland (www.animaland .org) is the ASPCA's interactive web site for kids who love animals. There's a lot to learn at Animaland! Regular features include kids and animals in the news, information about careers working with animals, fun activities, Animal ABCs, and Ask Azula, where ASPCA experts answer kids' questions about animals.

Do Something (www.dosomething.org/ aspca) is an online site that gives teens and tweens who want to make a difference lots of information and ideas about how to get involved in issues that matter to them. The ASPCA is the animal welfare partner of Do Something and gives $5,000 worth of grants each year to help fund worthy projects.

Index

Attention animal lovers!

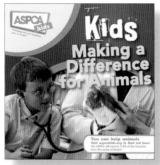

978-0-470-41086-8

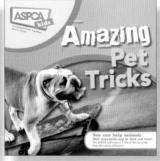

978-0-470-41083-7

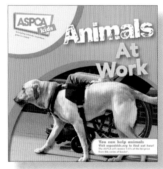

978-0-470-41084-4

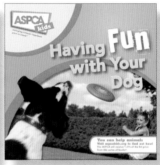

978-0-470-41085-1

Check out all of the ASPCA® Kids books to discover more fun facts about your favorite animals.

Available now wherever books are sold.
howellbookhouse.com

HBH Howell Book House™
An Imprint of WILEY
Now you know.